Think abou

game you have learned. Did your mother or father teach it to you? Maybe your brother or best friend showed you how to play it. Even games that are new to you could be old.

Some of the games we play today are like those that people in other places played thousands of years ago. Rules may have changed over time, but most are played in the same old way.

Read on! You'll find out about games from three different countries—China, America, and Mexico. You might want to try these games on a quiet summer day.

# A Chinese Game

**You Need:**

2 players

3 pieces for each player, with a different color for each

**To Play:**

**1.** Take turns putting a piece down on any open point on the game board.

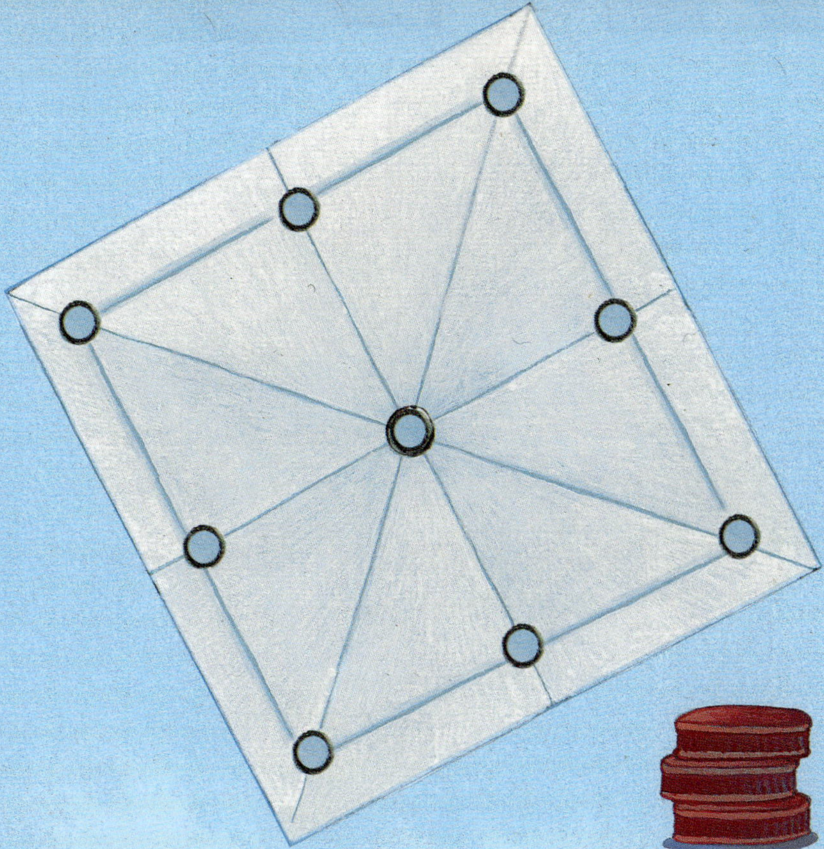

2. After all pieces are set, start moving one piece each turn. Push your pieces along lines to their next open points.
3. Study the game board. Look for ways to get three pieces in a row. If you can do that, you'll win.

# A Native American Game

**You Need:**

2 players

12 red pieces and 12 black pieces

**To Play:**

1. Set up the game board as shown.
2. Move one piece at a time by following any line to the next point.
3. Just as in a game of checkers, you can jump over another player's piece if there's an open point on its other side.
   - When a player jumps another player's piece, he or she takes that piece.
   - It's okay to jump more than once in a turn.
   - Players can change direction as they make several jumps back and forth.
4. You've won when the other player has nothing left.

# A Mexican Hunt Game

**You Need:**

2 players

1 white piece, 12 black pieces

**To Play:**

**1.** First cover the spaces on the board.
- The white piece represents the hare. Set that piece in the center space.
- The black pieces are coyotes. Place them on rows one and two and on the end points of row three.

**2.** Go one step each turn to a new open point.
- Only the hare can jump other pieces.
- When the hare jumps a coyote, it takes that black piece off the board.

**3.** To win the Mexican Hunt Game:

- Coyotes finally win when they stop the hare from moving.
- If a hare has caught enough black pieces, it becomes certain that the coyotes will never put it in danger again. That's how this small animal wins.

Now, if you can, invite a classmate to help you make one of these games, and then play it. Or, try all three! Once you know the rules, teach someone else how to play—for summer fun.